JUBILAT

By the same author:

The Necklace (1955)
Seeing Is Believing (1958, 1960)
A Peopled Landscape (1963)
American Scenes and Other Poems (1966)
The Way of a World (1969)
Written on Water (1972)
The Way In and Other Poems (1974)
Selected Poems (1978)
The Shaft (1978)
The Flood (1981)
Translations (1983)
Notes from New York and Other Poems (1984)
Collected Poems (1985; revised 1987)
The Return (1987)
Annunciations (1989)
The Door in the Wall (1992)

JUBILATION

Charles Tomlinson

Oxford   New York
OXFORD UNIVERSITY PRESS
1995

Oxford University Press, Walton Street, Oxford OX2 6DP
Oxford New York
Athens Auckland Bangkok Bombay
Calcutta Cape Town Dar es Salaam Delhi
Florence Hong Kong Istanbul Karachi
Kuala Lumpur Madras Madrid Melbourne
Mexico City Nairobi Paris Singapore
Taipei Tokyo Toronto
and associated companies in
Berlin Ibadan

Oxford is a trade mark of Oxford University Press

© Charles Tomlinson 1995

First published in Oxford Poets
as an Oxford University Press paperback 1995

All rights reserved. No part of this publication may be reproduced,
stored in a retrieval system, or transmitted, in any form or by any means,
without the prior permission in writing of Oxford University Press.
Within the UK, exceptions are allowed in respect of any fair dealing for the
purpose of research or private study, or criticism or review, as permitted
under the Copyright, Designs and Patents Act, 1988, or in the case of
reprographic reproduction in accordance with the terms of the licences
issued by the Copyright Licensing Agency. Enquiries concerning
reproduction outside these terms and in other countries should be
sent to the Rights Department, Oxford University Press,
at the address above.

This book is sold subject to the condition that it shall not, by way
of trade or otherwise, be lent, re-sold, hired out or otherwise circulated
without the publisher's prior consent in any form of binding or cover
other than that in which it is published and without a similar condition
including this condition being imposed on the subsequent purchaser.

British Library Cataloguing-in-Publication Data
Data available

Library of Congress Cataloging-in-Publication Data
Tomlinson, Charles, 1927–
Jubilation/Charles Tomlinson.
p. cm.—(Oxford poets)
I. Title. II. Series.
PB6039.0349783 1995 821'.914—dc20 94-30770

ISBN 0-19-282451-1

1 3 5 7 9 10 8 6 4 2

Typeset by Rowland Phototypesetting Ltd
Printed in Hong Kong

To Brenda

ACKNOWLEDGEMENTS

Acknowledgements are due to the editors of the following: *Agenda, American Poetry Review, Bête Noire, Critical Quarterly, Equivalencias, La Huella del Ciervo, The London Magazine, New Writing, Paradiso, Paris Review, Partisan Review, Poems for Alan Hancox, Poetry Nation Review, Scripsi, La Semana, Sette Poesie, Stand, The Swansea Review, The Times Literary Supplement, Triquarterly.* These poems first appeared in *The Hudson Review*: 'On the Terrace', 'To Be Read Later On', 'After Hugo', 'Pavane for a Live Infanta', 'To My Daughter', 'Autumn', 'Transaction at Mallards Pike'.

CONTENTS

For a Granddaughter
  1 On the Terrace — 9
  2 To Be Read Later On — 10
  3 Jessica Learned to Kiss — 11
  4 After Hugo — 12
  5 Pavane for a Live Infanta — 13
  6 To My Daughter — 15
Against Travel — 16
The Cypresses — 17
Lago di Como — 19
Varenna — 20
Asolo — 21
To Vasko Popa in Rome — 22
San Carlo ai Catinari — 23
Roman Fugue — 24
Gutenberg and the Grapes — 25
Horizons — 26
Valestrieri — 27
Down from Colonnata — 28
A Retrospect: 1951–91 — 29
Portuguese Pieces
  1 Alto Minho — 30
  2 Ponte de Lima — 30
  3 Soajo — 31
  4 Swallows — 31
  5 Oporto: St John's Eve — 32
  6 In Lisbon — 32
  7 Tagus, Farewell — 33
Zipangu
  1 The Pines at Hakone — 34
  2 Heron — 35
  3 Shugakuin Garden — 36
  4 Yamadera — 36
  5 Epilogue — 38

Interior	40
Snapshot	41
The Improvement	42
To A Yorkshireman in Devon	43
Crossing the Moor	45
Hay Tor	46
Jubilación	47
Durham in March	49
Near Bickering	50
Autumn	51
Weather Report	52
Cosmology	53
December	54
The Track of the Deer	54
To a Photographer	55
The Shadow	56
Walks	57
Transaction at Mallards Pike	58
The Song of Adam	59
To a Christian Concerning Ivor Gurney	60
9 a.m.	61
Knowledge	62
On the Late Plane	63

FOR A GRANDDAUGHTER

1 ON THE TERRACE
'Blest the infant Babe . . .'

Four of the generations are taking tea,
Except that one of them is taking milk:
It is an English, autumnal afternoon,
The texture of the air half serge, half silk.

It is an English, autumnal afternoon,
And all four of the seasons are sitting here,
Except that one of them lies interfused
With the flesh that feeds, the arm that cradles her.

The seasons are talking in a fugue of voices,
Except that she is trying out the sounds
Through which her tongue must learn to reach the words
To speak with the world which summons and surrounds

Her kindling senses: the circumference
Of many circles draws her from her warm
Dark continuity with all things close,
To know more than the flesh, the food, the arm—

That circle within the talking circle here,
By the old house, its stone-flagged passageway,
Within the circle of the lawn, the flowers, the trees,
The young attention widening where they sway.

2 TO BE READ LATER ON

Poets, my dear, are much the same as you:
Watching whatever shapes come into view,
They try a murmur, a melodious sound
To suit the sense of what it is they've found
And go on finding, as they write and pause,
Their aim as much the wonder as the cause.
I watched today what would have pleased you, too—
The shadows on the curtains where they blew
In at the window, shadows that showed how
(The frame quite rigid, yet the lines one flow)
Wrought metal can turn molten in the sun
Leaping along the muslins as they run,
A whirl of lattices, a flying net,
The whole breath of the day caught up in it.
Mallarmé (a poet you must read)
Wrote of une dentelle abolie—indeed
The sun writes on a curtain and erases
(In going out) those lines that are its phrases.
Whether the conflict is a birth aborted,
A Work unconsummated he had courted
(You'll spot the allusion when you read his verse
Or hear how Boulez makes the terse more terse),
For us that flowing through the window space
Could only fill more full the blowing lace,
As if our futures—yes, both mine and yours—
Were breathed towards us off the Severn's shores,
And when you lift this poem to your ear
One day, it is that breath of ocean you will hear.

3 JESSICA LEARNED TO KISS

Jessica learned to kiss,
Yet never would
Kiss me. This

Withholding of a kiss
Seemed to be
Part of her glee
At parting.

Or was she
Wise enough to see
That to defer
Made time doubt
Its hold on her
And me?

At all events
Only this week,
Perhaps disenchanted
With philosophic teasing,
A kiss she planted
On my cheek.

4 AFTER HUGO
'Jeanne songeait . . .'

She was dreaming, sitting on the grass:
Her cheek was pink, her gaze was grave:
'Is there anything you would like to have?'
(I try to anticipate her least desire
To find what it is that sets her thoughts on fire.)
'Some animals,' she said—just that, no more:
I showed her an ant in the grass—'There you are!'
But her imagination was left half-fed:
'No. Real animals are big,' she said.
Children dream of the vast, the ocean draws,
Cradles and calms them on the shore
With its rough music; its shadowiness
Will wholly captivate a child's mind,
And so will the terrifying flight of the sea-wind.
They love to be terrified, need wonder, feel no distress.
'I have no handy elephant,' I replied,
'So would you like something else?—just say what.'
Pointing a small finger at the sky, she answered 'That!'
Evening would be overtaking the world soon—
I saw climbing up above the horizon an immense moon.

5 PAVANE FOR A LIVE INFANTA

How would you like to wear
a yellow dress
such as this
in Velázquez's picture?
Your brocade
will make you feel at first
a little staid
perhaps, and so
you must learn
to dance the pavane, a slow
dance in which
your skirts will sail
over the sea-floor
of this giant palace
as noiselessly
as a painted ship.
The painter pauses:
he knows that he
can render only
your presence here
but not your movement.
Your favourite dog
has the air
of sitting for his portrait,
but has fallen asleep,
which only goes to show
how peaceful and how slow
a pavane can be,
and how skilfully
you dance it, gliding
past his nose
unespied. Your meninas
—your waiting maids—
cluster round
to slow more and more
your pace: they
and your dwarves
Maribarbola and Nicolás
think that your gait must grow

ever more sedate,
so that the artist can
at last proceed
to contemplate
you of the pavane
in utter stillness.

6 TO MY DAUGHTER

'Families,' I said, conscious that I could not find
    The adequate epithet, 'are nice.'
'Nice families,' you replied, adding
    To the faded adjective a tiny spice,
'Are nice.' What I had meant was this:
    How far we (a wandering family) have come
Since that day I backpacked you down
    Into an Arizona canyon with its river
Idling below us, broad and slow;
    Next, it was the steady Susquehanna;
Now swifter currents of the Severn show
    That time is never at a stand, although the daughter
You are leading by the hand, to me
    Seems that same child cradled in Arizona.
No—you are right—: *nice* will never do:
    But it is only families can review
Time in this way—the ties of blood
    Rooting us in place, not like the unmoving trees,
And yet, as subject to earth, water, time
    As they, our stay and story linked in rhyme.

AGAINST TRAVEL

These days are best when one goes nowhere,
The house a reservoir of quiet change,
The creak of furniture, the window panes
Brushed by the half-rhymes of activities
That do not quite declare what thing it was
Gave rise to them outside. The colours, even,
Accord with the tenor of the day—yes, 'grey'
You will hear reported of the weather,
But what a grey, in which the tinges hover,
About to catch, although they still hold back
The blaze that's in them should the sun appear,
And yet it does not. Then the window pane
With a tremor of glass acknowledges
The distant boom of a departing plane.

THE CYPRESSES

The cypresses are hesitating whether to move,
as though they could advance uphill if only they wished.
Then they go completely still; they are shamming dead—
they feel something is about to occur
and they want to be unnoticed by it.
Suddenly, we learn what it is:
the lake below, having lain in a Götterdämmerung light all
                                          afternoon,
disappears beneath cloud and rain. Now
the cypresses are losing their composure, but only a little.
They do not toss to the Byronic thunder music—
their foliage is too compact for that;
the brushed-up look of this leafy chevelure
has an electrical restraint about it.
Only a stirring at their very tips gives them away,
though the two just outside the window
begin conferring together, then one of them
perceptibly shudders through its entire length: rain
has arrived here too and the full force of it takes them
and—yes—twists and pummels and pushes them out of the true—
they who were the only plumblines on this uneven slope,
and the sense of the vertical goes out in a blackness
that might have been drained from them. The rain and the
                                          lightning
show that the unsubjective south does not start here: you are free
to read into this sunless ferocity
the presence of a tortuous god who hides his intentions
in smoke, and hints at them with explosions of light
right down to the bottom of this seething pit. Then the rain
climbs off, the trees come into focus once more
momentarily clear, and through the gaps in the foliage
you can see the lake again across which a ferry is passing:
it enters a gap and leaves it (it takes
five seconds for a craft to sail through a cypress tree).
Even the gulls are in circulation once more,
going round and round, stark white above the still-dark lake.
The cypresses overcome their uncertainty—this time
they are not going to be recognized—and emerge in disguise:
their clumps and companies have evidently

turned into a convocation of tall, thin clerics
poised at the foot of the incline on their way to a shrine
somewhere up at the summit. The pair of trees nearer at hand—
a tall one beside a smaller—are a mother and child,
the mother gigantic and the child likely to take after her,
though keeping its distance while listening intently upwards.
Thunder still resounds through the mountains
and the convocation has not yet moved away.
The ferry boat is re-establishing the timetable of the everyday:
Bellagio, Menaggio, Varenna, Bellano . . .
The shrine is catching the last light now
or is it merely an outcrop of white rock just by the peak
needing no further miracle or shaping story
to be what it is?

LAGO DI COMO

Sun must first filter through a haze
    That eats whole mountains here. How lonely
Those mountains would be without our presence.
    For only we can tell back to them their surfaces,
Their whites which absorb so many shades,
    Those surfaces accepting so much shadow
Into their clefts and crevices, yet marbling with light
    The lake beneath them, so that mountains stand
On columns that cross the waters, columns
    That undulate in flakes of white and gold.
This is the story told back to the mountains
    And, as evening begins, must be re-told,
To these summits taken by white fire
    From a sun that is already going, has not yet
Gone down behind the further shore whose crags
    Are climbing blackly upwards into silhouette.

VARENNA

Waiting for the ferry
we watch the late sun
gilding inordinately
the lakeside town

and the lake itself,
as if to insist that we
need look no further:
immanence is mystery

where the column of sundown
reflects in a vertical
tall encroachment
like flickering oil,

shoots into the harbour
flames that wrestle and dance
through each undulation
of the travelling substance,

till a peak comes between
and the fire-threads fray
and the darkening water
ferries night through the bay.

ASOLO
for Rosa Scapin

Fountains of limestone, limestone colonnades
    Reverberating like wells, porticoes of shade
And against the sky the dark of cypresses:
    Browning brought them back from Tuscany
To stand against the sunset. Terraces
    Could not mark the gradation of a hill
More exactly than their ranks, and when the sun
    Climbs down behind them, one
By one they offer it their stair
    To steady its descent then disappear.

TO VASKO POPA IN ROME

'Rome I dislike,' you said in French,
    'With its imperial pretensions.' You
Were the least imperious of men, in verse
    And person. We met only once again
And it was clear your days were near their end,
    Your life and death feeding on cigarette
On cigarette. 'Like a prince in exile,'
    Someone said, but that seemed fanciful for a man
Indifferent to empire. You were in exile from yourself,
    From that puzzled ebullience, watchful irony,
Balanced, it seemed almost bodily—
    For you were then a man of ample flesh—
Between Gallic precision, Italian largesse,
    As our conversation veered from tongue to tongue
In search of words adequate to express
    Our sense of the occasion. As to princeliness, I recall
Hearing you muse, 'Hughes, they say,'
    (Crossing the Borghese park near midnight)
'Lives like a prince.' 'That's true,'
    Was my reply, 'if generosity's what they intend,
And if you are his guest or friend, it's you
    Who live like one.' Pacing on,
Complaining of the melancholy great cities breed,
    As if all generosity must feed that, too,
You drew your gloom from a reserve of riches
    That soon must fail. In Rome, today,
I almost persuade myself you would agree
    That the bounty of the place exceeds pretension,
Bursting on one, as when the roar
    Of the Trevi fountain rounds the corner of its square;
And that these levels of wrought stone and water—
    Metamorphosis over an ungiving ground—
Are one more form of poetry, and we
    Guests of the imagination here. The imagination
Proposes what it does not need to prove
    And, when all's said and done, what cannot be:
Now we shall never pace this square together
    Through the Roman sunlight and the autumn air.

SAN CARLO AI CATINARI

An angel orchestra
have just fluttered down
in stone and settled on the rim
of the dome to hymn
Saint Cecilia.

I admire this scene
so far above my head
for its solidity: these
are no shadowy presences
but flesh and stone interinanimated.

If we were angels
we could no doubt hear
their silent music
fleshed in the substance
of another sphere—

A sphere that sense
enters but rarely and when it does,
gathers more palpable evidence
of what it is
so delights it here.

For what could heaven imply
but the increase and care
of each tuned faculty
turning to attend and praise,
imaged in that high consort there.

ROMAN FUGUE

Beyond the window
from their rooftop terrace
the backs of three
statuary busts
gazing apparently
towards a striped awning
builders have draped
before their operation
on a pilastered
and arched façade: half
a builder—cut in two
by that sheet or screen—
walks an unseen plank
intently searching
for something just out of view
ignored by the procession
above his head
of small clouds crossing the Roman blue
in contrary motion.

GUTENBERG AND THE GRAPES
for Bill Murphy

Watching them turn the screw
tighter, tighter
above the press
he knew at last
what it was
he was looking for:
it must have been
the winey air
had opened his senses
and imprinted the secret there
as the great block
bit into pulp below
and the wine-clock clink
of the machine
ticked towards a time
beyond him like the strokes
on an anvil: the copiousness
of the grape was filling
the cask before his eyes:
wine had gone to press.

HORIZONS

for Bruna Dell'Agnese

'That imaginary, uncrossable line we call
    Horizon': it is, of course, illusion
Like so much else, and yet our eyes
    In their myopia surprise the truth. I stand
Where a hill above the cliff takes in
    Not one but two horizons—land
And, as if the deluge hung in momentary abeyance,
    The blue curvature of the sea above it,
Impending with its weight of liquid acres.
    But, no, what we carry from the scene
Is less the image's apocalyptic threat
    Than that parallel measure, as intangible
And clear a presence as the spanning rainbow,
    That goes on telling us the world is there
And what shape it has, the bow itself
    Nothing but sunlight, water, air.
That imaginary, uncrossable line
    Confronts us with as fine a demarcation
As the sailing, selvedged clouds
    Balancing along the wind and above the sea
Their changing images true and imaginary.

VALESTRIERI

for Astrid Donadini

The bridal veils of the olive trees, you said,
    Seeing the white nets spread
Underneath the boughs. But these
    Slung higher to catch the crop
In its fall are the hammocks of the autumn voyage
    Into winter, swung in the after-gale
That follows the first bright cold
    Cutting mist, bringing back sun
Into the orchards here. They have cleared the ground
    Of its brush where the nets must lie—
There is to be no waste—and all is readied
    For the slow maturing of berries still green.
But the echo of volleys through the colder air
    Bursts from the presence of huntsmen there, unseen,
Lying low claiming consummation now
    In the pattering ricochet of aim on spendthrift aim.

DOWN FROM COLONNATA

A mist keeps pushing between the peaks
    Of the serrated mountains, like the dust
Off marble from the workings underneath:
    Down from Colonnata you can hear
The quarrymen calling through the caves
    Above the reverberation of their gear
Eating through limestone. We are moving
    And so is the sun: at each angle
Of the descending road, the low light
    Meeting our eyes, surprises them whenever
It reappears striking a more vivid white
    From the crests behind us. Down
And on: the distance flashes up at us
    The flowing mercury of the sea below
That we, passing Carrara, lose
    Until it shows once more backing the plain.
But the sun has outdistanced us already,
    And reaching the level water, dipped
Beneath it, leaving a spread sheen
    Under the final height dividing us,
And across the liquid radiance there,
    A palpitation of even, marble light.

A RETROSPECT: 1951–91

We go down by the deserted mule-track:
    Myrtle berries and purple daisies overhang
This unused pathway of cracked stones
    The walls wind round with. It leads
Between netted olive trees and enters
    La Serra from above, down past the house
Its poet was born in, that will one day—
    This is a country of inscriptions—bear
Let into its wall, a crisp-cut *lapide*.
    You could still hear his mother tongue
(His mother's tongue) if only you
    Could speak it and could call out to
That woman who descends in front of us,
    Her kindling carried on her head as when
We first came here to streets that have withstood
    Corsair and scimitar. A poor place then,
But its stone severity hospitable
    With wine and conversation round a fire
That stung the eyes with woodsmoke. She
    That solid apparition, has disappeared
Along her alley, as we turn to cross the square,
    Into a rawness blowing off the bay
That tells how the season and the world
    Are travelling to where these forty years began
In a tumultuous autumn of seastorm, cold and rain.

l. 7: the poet is Paolo Bertolani, author of the dialect poems, *Seina*.

PORTUGUESE PIECES

for Gualter and Ana Maria Cunha

1 ALTO MINHO

Não, não é nesse lago entre rochedos...
                         Pessoa

Bees move between the rosemary and the rose.
The oranges are waiting to be picked.
The coigns of granite by the threshing floor,
The inscription of the runic mason's mark
Ask to be clarified by the hidden sun.
(Later, it will break along the river
To show where the waters of the floodtide reached
And stained with mud the lower leaves of trees
The colour of stone, a petrine fringe reflecting
In the calm beneath . . .)
Here, bread and reality are reconciled
By the excellence of maize, the spread hunks.
We are eating honey in a granite house.

Quinta do Baganheiro

2 PONTE DE LIMA

Lima was *limes*, limit—
beyond the river, only the mountains.
On its bank, the alameda of tall plane trees, now,
and ghostly washing
that catches the final light, the flow
of still-warm air. The blade
of the river is broken
by the housetops and the trunks
that rise between the eye and it
on the brink of the unimaginable,
its sinuosities unclear.
Was it *limes* or *limen*, limit or threshold?
They called it Lethe, the Romans, and bridged it.
Their bridge is still here.

3 SOAJO

A glitter of particles
embedded in bedrock—
no asphalt here: a jigsaw of granite
paves the village square.
Granite curves the well-kerb,
granite guards the grain:
from a dais of staddlestones
looms a mausoleum for maize
that rings the hill-top
with tombs for a dynasty of kings.

4 SWALLOWS

Swallows outshout
the turbulent street:
swallows are messengers
where the day and night meet,
bringing news
from gods older than those
who pose in the gold interiors,
on the tiled cloister wall;
and a swallow it was
that arrowed past
threatening to graze you,
but delivered itself instead,
disappeared into
the dark slot above
a lintelled doorhead.

5 OPORTO: ST JOHN'S EVE

At this pagan festival of St John,
Churchkeys click in the shut lock.
Gilt and silver glint from the resonant dark.
Their surfaces are playing with the fire
Youth is leaping through outside.
The iron parapet
Sways with the crowd's weight
Above the river. In hillslope spate
Sheer saturnalia flows.
Are those salutations blessings or hammer blows?
The midnight churches are looking the other way,
Like the public plinths
Where sculpted deeds are done,
And the bedraggled Eagle still cedes to Wellington.

6 IN LISBON

At the Versailles
The waiter talks of his pride in the place,
With its ornate soffits, mirrors, glistening wood.
Once, he had gone to look upon the face
Of the real Versailles
To see how the two compared, and found
The ceilings in the apartments of the Pompadour
Were just like these . . . Pessoa, all around,
Demolitions are dragging your city down,
And cranes constructing the blank bank architecture
The future will know us by. At the Versailles
We reconsider the Pompadour and find
Only by style will you engage the affections of the mind.

7 TAGUS, FAREWELL

It is a very filtered light
    Permits this fine gradation of the fields
From the passing train. It is the cloud decides
    The softness of these shades of green. This
Is an English summer scene renewed
    To eyes returning from Iberia's blaze
On glittering granite: Tagus, farewell.
    When Wyatt came, with spur and sail,
Back into Kent and Christendom, and found
    Thames like a bent moon-bow,
The river running through English ground
    Exposed 'her lusty side', mistress
Or doubtful bride. So she remains
    In this same world of whim, of trade and trains.
But the light deals no deceit that sees
    The same month ripening now that brought
Wyatt to England and unease, a mind in woe,
    Closed to the sweet complexities of weather.

ZIPANGU

for Yoshiko Asano

1 THE PINES AT HAKONE

The pine trees will not converse with foreigners. Their aim
is to hide everything that lies beneath their crisp, dense foliage
or at their feet—those ferns, for instance, that reproduce
the pine pattern on every leaf and lie low
the air scarcely stirring them. They have learned
to keep secrets by studying the tall trunks that surround them
and that might still be living in the Edo period.
Touched by the breeze, they rock on their pliant roots
and shift slightly their green vestments, beginning to oscillate,
to lean from side to side, even to bow—
though not deeply as is customary with this people—,
as if good manners were all they had on their minds
and they had spent a long time considering the question
without coming to any conclusion. The tiny agitations of the wood
are on the surface only, and they soon resolve themselves
into the general unison of branches, heaving, subsiding.
Today the clouds are as secretive as those branches
and they refuse to reveal the summit or the sides
of Fujiyama. You sense it there, but you cannot see
its bulk or its snow-streaks that Japanese art
has made so famous. (Hiroshige was here
but on a clearer morning.) Days later
and back in the capital, I watch the carp
in the pond of the Yasukuni shrine. These fish
in their extrovert muscularity, their passion for food
are all the trees are not; they steer themselves unerringly
with a blunt muscular force, their whiskered circular mouths
forming the O which means *give*, rolling over
on one another's backs, to get what is given, and arriving
with the massive bodily impetus
of legless sumo wrestlers ruddered by flickering tails.
But this is a military shrine, its gate a tall ideogram
topped by a bar like a gigantic gun-barrel
and the mere good manners of trees do not serve here
to distract the visitor from what he wishes to understand.

Though when he rises to go, the lit lanterns,
as if disguise were after all the mark of this nation,
throw through the branches a light of festivity, a carnival glow,
their object solely to beautify the spot
and make us forget what stern ghosts linger here.

2 HERON

The river crosses the city over a series of falls:
at each of the falls, waiting for fish
a small white heron—sometimes
a whole group of them, all
at a respectful distance from one another.
Perhaps they have fished the river too long—
they seldom visibly produce anything from it.
Perhaps their decorative tininess is the result
merely of malnutrition. They are indifferent to traffic
flowing by on either side, and to strollers
who pause to see what they might catch.
They watch the water with such an exemplary patience,
they seem to be leftovers from a time
when the world was filled with moral admonitions
and everything had been put there to mean something.
We, however, fail to take the lesson to heart
and continue to worry over their inadequate diet.
As evening arrives, the light on the buildings
goes golden, an Italian light, and the mountains
darken and press forward to stand protectively
round the city. Midnight
and below the roadway, in the glare of passing cars,
huddle the heron, roosting with one leg raised, and bent
even in sleep, towards the flash, the fish, the disappointment.

Kyoto

3 SHUGAKUIN GARDEN

The variegated tremor
of the reflected foliage
brings autumn to the ponds:

the rising fish
create circles within circles,
pools within pools:

under garden branches
there is every sort of water
to be seen and listened to

as it talks its way downhill
through the leat of its channel
out into the rice beyond:

you will find no frontier
between the garden and the field,
between utility and beauty here.

4 YAMADERA

You go by the local line:
schoolchildren keep getting off the train,
returning to those villages
beneath vertical mountains.
Kumagane: conical hills
beyond the little station;
Sakunami: the sky is darkening
and so are the trees;
it will rain soon—in time
for our arrival
by this narrow way
to the deep north,
though 'deep', they say,
is a mistranslation

of the title of Bashō's book,
and 'far' would be more accurate
'though less poetic,' they add.
The river in the ravine,
this intimate progress between sheer slopes,
what must it have been
for a traveller
on foot and horse-back?
Our rail-track way
is a smooth ascent
through turning maples
into cooling autumn air,
the faint aroma of snow in it.
It was here he wrote—
but would not write today—
the shriek of the cicada
penetrates
the heart of the rock.
He came, then, in heat.
The climb up the mountain face
which is the temple
must have cost him sweat,
his feet on the thousand steps
that lead past the door of each shrine
up to the look-out where
you can take in the entire valley,
echoing, this afternoon,
with shot on shot
from a whole army
of automatic scarecrows.
The rain arrives, but does not stay,
from a grey cloud
darkening half the sky
and disappearing. On the way down
we see once again
what arrested our upward climb—
stones to the miscarried,
and prayer-wheels
to wish the unborn
a reincarnation in a human form.
And so we depart
in the light that saw his arrival—
that of late afternoon,
to wait for the train

in this still distant corner. Clearly
the poetry of 'deep'
is more accurate
than mere accuracy—
a journey to the interior
is what it must have felt like then.
They say he came as a spy
(the villages are passing in reverse order now),
that there was more to it than met the eye,
calling on abbots and warriors,
to sniff out plots before they occurred.
There is no doubt, some say,
others that it is absurd
to speculate now. And so
we leave Bashō to disappear,
deeper and deeper,
while we cross the angular paddies
towards the shapeless cities,
the mountains already drawing apart
on either side of the wide plain
into two great parallels
echoing the track of our train,
our own narrow way south.

The title of Bashō's travel book, *The Narrow Way to the Deep North*, is the translation (or invention) of Nobuyuki Yuasa (Penguin Books, 1965).

5 EPILOGUE

This advanced frontier
of Asia, this chain
of volcanoes, arcadian,
alpine, weird,
its ravines noisy with waterfalls,
its countless rivers
too impetuous for navigation,
ports few and coast foam-fringed—
the tree-fern, bamboo,
banana and palm grow here
side by side

with pine, oak, beech and conifer.
Wild animals are not numerous
and no true wolf exists
(the domestic dog
is wolf-like but ill-conditioned).
The lobster stands for longevity
and all history before 500
must be classed as legendary.
This is the place
Marco Polo never visited
but, jailed by the Genovese,
rehearsed its wonders
in bad French
to a Pisan fellow prisoner
calling it
Zipangu.

INTERIOR

Approaching the house,
he lingers at the door;
it is the thing he sees inside
delays him there:

the drawing-room
seems to essentialize
asters, dahlias and golden-rod;
the red-gold sunlight lies

in puddles across the floor,
turns the blue of the carpet
more blue yet and leaves
a hazy aureole round each chair:

the colours of autumn
within four walls burn
more richly than maples, asters
fanfaring his return

to this domestic fire
carried indoors from outside:
here is a hearth rekindled
from the whole red tide

by the reconnaissant hand
that hesitating must pause
before it places its final flower,
stands back from it and withdraws.

SNAPSHOT
for Yoshikazu Uehata

Your camera
has caught it all, the lit
angle where ceiling and wall
create their corner, the flame
in the grate, the light
down the window frame
and along the hair
of the girl seated there, her face
not quite in focus—that
is as it should be, too,
for, once seen, Eden
is in flight from you, and yet
you have set it down complete
with the asymmetries
of journal, cushion, cup,
all we might then have missed
in that gone moment when
we were living it.

THE IMPROVEMENT

            The hallway once
    Ran straight through the house, and you could see
Entrance to exit in one sweep
    Of the eye across a cobbled floor—
Unexplored, the territory of the rooms
    To either side. One day
It was resolved to block that shaft
    With a vestibule, and to curb its tendency
With a further jut of wall half way.
    Do I like the changes? A ghost
Could not pass straight through
    Without confusion now. What ghost, you say?
The ghost that is my memory which quickens still
    At the thought of the long passage lit
From door to door, the clean
    Flight of the senses through it, like a wife
Running to meet her man, like a bird's flight, a life.

TO A YORKSHIREMAN IN DEVON
for Donald Davie's 70th birthday

Eden was never Abyssinian—
In spite of Milton and received opinion.
He chose, at last, the paradise within.
'Within', without the rest of it, sounds thin—
Even to one like me who, as you say,
Has gilded rural scenes inordinately:
I could not live only on leaves and grass
For all my equanimity, but let that pass.
Gurney thanked God for Gloucestershire. You see
At once how a mere county boundary
Could not explain the intentions of the Lord.
And yet, is it, Donald, utterly absurd
Like Edward Thomas to accept a war
Convinced it was Eden you were fighting for?—
That Eden Gurney found on midnight walks
Glimmering along boughs, up nettle stalks,
Through constellations that the Romans knew
Standing in that same damp of Cotswold dew
On sentry go. And Gurney's thanks began
With the Georgic feelings of the Englishman
For land that is worked. And so his Eden means
The practicality of rural scenes
Besides the poetry of place—divine
And human, not too rigorous a line
Severing the two creations. I,
A gardener beneath our doubtful sky
Hoeing my beans—no, not nine rows like Yeats
(His were the kind one neither plants nor eats)—
See you beside your lawn of camomile
(In thought, that is—visions are not my style),
With pipe and books and mollifying glass,
Challenge the ill-kempt verse that tries to pass
The approbation of your level gaze,
Though not so partial that you cannot praise
Writers whose premises dispute your own,
Oppen and Olson, Niedecker and Dorn—
Gurney himself whom we rejoice to see
With Bunting at our island's apogee.

So if I must decide on qualities
That show you as you are, it's my surprise
At where you'll lead us next that makes my task
So difficult a judgment, when I ask
What are the limits within such a mind
That's principled, yet never is inclined
To set up and defend impossible frontiers,
Brooding on words and meanings these long years;
And now your great climacteric's wide south,
A region to delight and nourish us both,
Offers beneath the shadow of the moor
The Sabine promise of your open door.

CROSSING THE MOOR
for Fred and Paula

Crossing the moor, the prehistoric stones
    Keep to their circles and their avenues, imprint
On earth a planetary map that we
    Can read no longer. We enter
The fractured ring; the veering weather
    Entering with us, brings out the glitter
From the grey: sunlight has made
    Malleable the solidities and so has shade
As the monoliths darken beneath cloud;
    Then a tongue of flame, flickering through the air
Leaves its glancing brightness everywhere
    Setting light to stone. The great
Clock of circle and avenue still
    Measures for us, chartless as we are,
Weather, space and time—the mobile features
    Continually re-forming of that face
The universe turns towards us where
    We cross by the sea-lanes of the open moor.

HAY TOR

The moor is starred with yellow tormentil
    In a rune of stones, as though some citadel,
Detonated here, had strewn the hill
    With fragments, and sheered away
From words to soundlessness—stone patiences
    Outwaiting time. Circles and avenues remain
To measure the naked ground, that then
    Were the forest markers, maps of processionals:
Landlocked among leafage, men
    Sought for the open circle and were led
Between trunks and growth where now
    Under a bare sky gorse and heather
Climb through the debris ivy holds together,
    And the constellations of the tormentil
Channel a wind, blowing as it lists,
    Through the breathing spaces where gone villages
Felt the stir of foliage fingering their stone.

JUBILACIÓN[1]
a letter to Juan Malpartida

You ask me what I'm doing, now I'm free—
Books, music and our garden occupy me.
All these pursuits I share (with whom you know)
For Eden always was a place for two.
But nothing is more boring than to hear
Of someone's paradise when you're not there.
Let me assure you, robbers, rain and rot
Are of a trinity that haunt this spot
So far from town, so close to naked nature,
Both vegetable and the human creature.
Having said that, now let me give a sample
Of how we make short northern days more ample.
We rise at dawn, breakfast, then walk a mile,
Greeting the early poachers with a smile
(For what is poetry itself but poaching—
Lying in wait to see what game will spring?).
Once back, we turn to music and we play
The two-piano version of some ballet,
Sacre du Printemps or Debussy's *Faune*,
On what we used to call the gramophone,
To keep the active blood still briskly moving
Until we go from dancing to improving
The muscles of the mind—'in different voices'
Reading a stretch of Proust, a tale of Joyce's.
And so to verse. Today, the game lies low,
And Brenda, passing, pauses at the window,
Raps on the pane, beckons me outside. She
Thinks, though we can't plant yet, we still can tidy,
Clear the detritus from the frosty ground
With freezing fingers, and construct a mound
Of weeds and wood, then coax it to a red
And roaring blaze—potash for each bed,
As Virgil of *The Georgics* might have said.
I signal back my depth of inspiration,
The piece I'm finishing for *Poetry Nation*
(What nation, as a nation, ever cared
A bad peseta or a dry goose turd
For poetry?). Our Shelley's right, of course,
You can't spur on a spavined Pegasus
Or, as he puts it, 'There's no man can say

I must, I will, I shall write poetry.'—
Or he can say it and no verse appear.
As you now see (or would if you were here)
The winter sunlight sends its invitation
To shelve these mysteries of inspiration
And breathe the air—daybreak at noon, it seems,
The swift de-misting of these British beams
(Our watercolour school was full of such
Transient effects—we took them from the Dutch).
Strange how this wooded valley, like a book
Open beneath the light, repays your look
With sentences, whole passages and pages
Where space, not words, 's the medium that assuages
The thirsty eye, syntactically solid,
Unlike the smog-smudged acres of Madrid
Boiling in sun and oil. You must excuse
These loose effusions of the patriot muse.
Not everybody's smitten with this spot—
When Chatwin lived here, he declared he was not,
His cool, blue eye alighting only on
Far distant vistas Patagonian,
Untrammelled in the ties of local life,
Lost to the county, to both friends and wife.
We'd walk together, talking distant parts—
He thought we all were nomads in our hearts.
Perhaps we are, but I prefer to go
And to return, a company of two.
Hence jubilation at my *jubilación*
That we, together, leave behind our *nación*
And visit yours—or, just look up, you'll see
The vapour trails above us, westerly
The high direction of their subtle line,
Spun between Severn and Hudson, and a sign
That we shall soon be passing at that height
And, if the weather's clear, catch our last sight
Of Gloucestershire beneath us as we go.
But I must use 'la pelle et le râteau'
(Things that were images for Baudelaire),
And with the backache, spade and rake, prepare
The soil to plant our crops in on returning.
So I must pause from versing and start burning,
To anticipate the time we're once more here
In the great cycle of the ceaseless year.

[1] The Spanish for retirement.

DURHAM IN MARCH

You can take it in at a glance
    This climbing town above its river,
Backed by the arches of a viaduct
    That ushers you in, prince of the place
It seems—prince-bishop, rather,
    Its castle and its coinage in your grasp.
For three cold days, your eyes
    Will rule over gulls and weirs,
Ramparts pierced by your posterns,
    The trinity of towers. Then time
Brings round the abdication of these powers.
    The train is moving. Below the parapet
Roofs and terraces, contouring-out the drop,
    Hold up their image, clear as on a coin,
Of a town compacted. Place, once more,
    Has outrun your words, your links of sound
Echoing back a sisterhood of shapes,
    Driven apart now by the growing space
Between you and them where they attend
    On the weathers that mark-off your three-day reign.

NEAR BICKERING

The seam that runs sinewing England
    Crops out here. I recognize
The colour of home, the Cotswold colour,
    Across these high wolds of the north,
In the free-stone barrier that divides
    Lawn from corn-field. An apple
Lies like a tiny boulder in the grass
    Bruised by its fall, and brown as stone;
Corn of the selfsame colour does not yield
    To the push of air. It is Yorkshire August here,
Cool even in the sun, and calm,
    So that only the thistledown can show
Where the air currents move, and drifts
    Earthwards with silk-thread tentacles
Reaching for rootage as they brush the ground,
    Aerial gleams through so much petrifaction,
First silent sorties in this truce with stone.

AUTUMN

                Neither pink nor brown,
    You cannot take it down in ink
The colour of fallen leaves—a brush
      Might match it, or music even
In an upward scale of infinite intervals
      With a change of key where a hill
Meets heaven and all the degrees
      Of autumn on ground, slope, trees
Clash suddenly with—no adjective
      Can define the blue that brightens
Through the whole sky and, in the chemistry of sight,
      Changes this glow that crackles at our feet
By being its antithesis: far, cold,
      Never to be trodden, that is where
Comparison ceases and we breathe the atmosphere.

WEATHER REPORT
for Brian Cox

First snow comes in on lorries from the north,
    Whitens their loads—an earnest of that threat
Cromarty, Mull, Fair Isle and Fasnet
    Have weathered already. It has passed
Down the Pennine chain and choked Shap Fell;
    The Snake is lost and every moor
In Derbyshire under a deep, advancing pile.
    It covers the county, dwindling south,
But the wind that carries it, overshoots
    The frontier snow has mapped. It is the wind
Seems to be blowing the sunlight out
    As it roams the length of the whole land,
Freezing the fingers of tillers and of trees,
    Until it curls back the tides off Cornwall
Telling the snowless shires they too must freeze,
    In this turbulence that began as Swedish air
And has turned in the translated atmosphere
    To the weather of the one nation we suddenly are.

COSMOLOGY

Where is the unseen spider that lets down
    This snow-web from the meshes of whose cold
Birds, stung out of stupor, suddenly unfold
    Into flight? The hungry huntsmen crows
Dismember the unmoving, beak them off bushes,
    Haling them up to their tree-high end.
There, strung with sinews of white the boughs
    Twist into a wickerwork from which a snow
Of feathers flows down as the shrieking ceases.
    The sun emerges now to show it all plain
And dazzle us with the same questions still.
    We pursue them anew with legendary solutions
Each dawn disowns, then leaves us standing here
    Rich orphans among an inheritance of snows, of stones,
Feeling the warmth ebb back into the year.

DECEMBER

Frost followed frost, each colder
     Whiter than the one before. The crystals'
Sparkling salt, it seemed, had changed
     The nature of the things it clung to:
You walked in a world that well might break
     Into glassy chimes, gamelans of the cold,
As whole hillsides, struck by the light,
     Stood out revealed, trees ranked in white,
Their detail microscopically incised, unreal,
     Sun full on their fragile armour that soon would
Melt off them in a single afternoon.

THE TRACK OF THE DEER

          . . . . . The track of the deer
That strayed last night into the garden,
Stops beneath the fruitless apple tree,
Shaped out and shimmering with that frost
You can feel here at the edge of all imaginings:
The departed deer glimmers with the presence
Of sensed, substantial and yet absent things.

TO A PHOTOGRAPHER
for Justine

The house in the paperweight is covered with snow:
The camera caught it in the boughs' embrasure,
And all is a circular window where the glass
Is keeping from time this moment of winter leisure—

Before the snowplough has opened the road again,
And the white on the roof has started to slide away.
The limewash lit by the sunlight seems a glow from within,
Like the pleasure of bodily warmth on a winter's day.

I can take up that day in my hand once more,
Now the springtime has altered this scene of your art,
But must wait out the year to breathe in its cool and find
How your image, like air, has entered and tempered each part.

THE SHADOW

The sun flung out at the foot of the tree
A perfect shadow on snow: we found that we
Were suddenly walking through this replica,
The arteries of this map of winter
Offering a hundred pathways up the hill
Too intricate to follow. We stood still
Among the complications of summit branches
Of a mid-field tree far from all other trees.
Or was it roots were opening through the white
An underworld thoroughfare towards daylight?
There stretched the silence of that dark frontier,
Ignoring the stir of the branches where
A wind was disturbing their quiet and
Rippled the floating shadow without sound
Like a current from beneath, as we strode through
And on into a world of untrodden snow,
The shadow all at once gone out as the sun withdrew.

WALKS

The walks of our age
are like the walks of our youth:
we turned then page on page
of a legible half-truth

where what was written
was trees, contours, pathways—
and what arose as we read them
half conversation, half praise—

and the canals, walls, fields
outside of the town
extended geographies
that were and were not our own

to the foot of the rocks
whose naked strata threw
their stone gaze down at us—
a look that we could not look through.

That gaze is on us now:
a more relenting scene
returns our words to us,
tells us that what we mean

cannot contain
half the dazzle and height
surrounding us here:
words put to flight,

the silence outweighs them
yet still leans to this page
to overhear what we talk of
in the walks of our age.

TRANSACTION AT MALLARDS PIKE

for Richard Verrall

The trunks of the spruce at Mallards Pike
    Float their reflections out across the lake
Into its depths. This columned church,
    This underwater shrine, sways in the wind;
Its dark recession might well be a mine
    Like those on either shore, but here
It is the light transforms itself to ore
    Not to be sold: the sliding darks
Yield up no miner's harvest of black gold
    To be weighed and traded afterwards.
Even the swimmer through that foundered nave
    Is robbed of the wealth of it we have
In our dealings with a sun and surface that
    Offers itself as mirror to the trees,
Then jostles their rigid tallnesses to a play
    Caught only here between the wood and light—
Walkers on water dark alone can drown,
    Weightlessly undulated between dawn and sundown.

Forest of Dean

THE SONG OF ADAM

It was the song of Adam
the devil envied most,
and the song of Adam
that Adam lost,

and could catch only
the attenuated echo
as he wove an accompaniment
to that remembered flow,

as if he might restore
from the tune in his head
the variable, self-sustained
unfaltering aria, fed

from the steady spate
and the hidden source
Adam and devil can only hear
as one leaden curse.

TO A CHRISTIAN CONCERNING IVOR GURNEY

You will have much to explain to your God on the final day,
And he, also, will have much to explain to you—
Why (say) the mind of Gurney, whose preludes I am listening to,
Should, through so many years, have to waste away
Into inconsequence—composer, poet who dreamed that our land
Would greet in him an heir of Jonson and Dowland;
But its mind was elsewhere, and so was that of your Lord,
Assigning this soldier his physical composition—
That blood, those chromosomes that drew him to the absurd
Disordering of notes, to the garrulity of the word,
Instead of the forms that already his youthful passion
Had prepared for the ordering of both self and nation.

9 A.M.

              For a long time
    There is sunlight in the sky, but in the streets no light;
Then it arrives and walkers going by
    Are joined to their shadows. Overhead
On a tall building a flag
    Twists shadowless in the morning air:
Why so, when there are shadows everywhere?
    I see a running man, dwarfed
By his shadow-legs; a man whose shadow-stick
    Trails a long stilt to far-off shadow-feet;
A pigeon flies down to land
    On its inverted cut-out, and the cars
Carry for a moment the shadow-shaft
    Of the lamps they are passing under. Now,
Screened far above me on a single
    Facet of sky-high wall, I see
The shadow-flag an hour will efface,
    Passing its serpentine black rag
From side to side across a surface
    Of resplendent concrete it is energetically cleansing.

KNOWLEDGE

I want the knowledge of afternoon as it recedes
    Up to the very edge of disappearance;
And of all the space I see it passing through,
    And all of the time it takes for light
To linger out its slow retreat from sight—
    Tree after tree, roof after roof—
And sensing them present there, to feel
    All that the charged recesses of the dark conceal.

ON THE LATE PLANE

The city is spreading its nets to catch the eye
    Of flight after flight crossing this night sky—
Nets of illumination that distance turns filigree,
    The fine-spun logic of an arterial geometry
Webbing outwards and on from each settled spot
    To reach to wherever the light still is not—
To pockets of dark where mountain sheernesses
    Ink-out the gossamer glow with their million fir trees.
Here is a star-map brought to the ground, a beauty
    Bestowed on the innumerable roadsides that by day
Harbour no hint of this transfiguration where
    They draw down the gaze of the night-sky traveller.

OXFORD POETS

Fleur Adcock	Jamie McKendrick
Moniza Alvi	Sean O'Brien
Kamau Brathwaite	Peter Porter
Joseph Brodsky	Craig Raine
Basil Bunting	Zsuzsa Rakovszky
Daniela Crăsnaru	Henry Reed
W. H. Davies	Christopher Reid
Michael Donaghy	Stephen Romer
Keith Douglas	Carole Satyamurti
D. J. Enright	Peter Scupham
Roy Fisher	Jo Shapcott
Ida Affleck Graves	Penelope Shuttle
Ivor Gurney	Anne Stevenson
David Harsent	George Szirtes
Gwen Harwood	Grete Tartler
Anthony Hecht	Edward Thomas
Zbigniew Herbert	Charles Tomlinson
Thomas Kinsella	Marina Tsvetaeva
Brad Leithauser	Chris Wallace-Crabbe
Derek Mahon	Hugo Williams